Cognitive Analysis of EEG Waveforms Using Neurofeedback Application

Clinical Psychological Case study

Cognitive Analysis of EEG Waveforms Using Neurofeedback Application

Clinical Psychological Case study

R.Kishore Kanna
R.Vasuki
N.Kripa
L.K.Hema
R.Gomalavalli

ELIVA PRESS

ISBN: 978-99949-8-717-7

Cover Design: Eliva Press
Cover Image: Ingimage
Email: info@elivapress.com
Website: www.elivapress.com

This Eliva Press imprint is published by the registered company Eliva Press Global Ltd. part of Eliva Press S.R.L. Publishing Group

The registered company address is: Pope Hennessy Street Level 2, Hennessy Tower Port Louis, Mauritius
Eliva Press S.R.L. Publishing Group address is: Bulevardul Moscova 21, Chisinau, Moldova, Europe

COGNITIVE ANALYSIS OF EEG WAVEFORMS BEFORE & AFTER PSYCHO TASK USING NEUROFEEDBACK APPLICATION

by

Dr. R. KISHORE KANNA*[1]

Department of Biomedical Engineering

Vels Institute of Science, Technology & Advanced Studies [VISTAS]

Chennai, India

kishorekanna007@gmail.com

Dr. R. VASUKI[2]

Department of Biomedical Engineering

Bharath Institute of Higher Education and Research

Chennai, India

Dr. R. KRIPA[3]

Department of Biomedical Engineering

Bharath Institute of Higher Education and Research

Chennai, India

Dr.L.K.Hema[4]

Department of Biomedical Engineering

Aarupadai Veedu Institute of Technology (AVIT),

Chennai, India

Dr. R Gomalavalli[5]

Department of ECE,

Siddharth Institute of Engineering and Technology

Chittoor, Andhra Pradesh, India.

ABSTRACT

Stress has a substantial positive impact on human performance. While little stress might help with attention retention, excessive stress reduces productivity. The severity of the issue with human stress levels is shown by this terrible number.

Individuals who reside in wealthy and more educated areas are more prone tohave these kinds of issues. Although relaxing techniques considerably lowerpupils' stress levels, cognitive responses to stress impair their capacity to focus.

Also, studies on PMBSR (Physical Mindfulness Based Stress Reduction) have demonstrated that it may lessen stress, chronic pain, and pain intensity. Yoga has physiologic health benefits including improved immune function and decreased arousal levels, as well as the potential to create brain alterations.

The study found that after practising yoga for the recommended amount of time, children, students, and working adults all increased their physical and cognitive strength.Based on the Pystack (go-nogo) test and pre-post yoga, the stress is determined. The electroencephalogram (EEG) equipment with 24 channels was used to capture the brain signal, and the Independent Compound Analysis (ICA) tool was used to filter out and classify the distracting sounds from the data.

Yoga stimulates the brain's prefrontal lobes, making the limbic system's awakening more challenging over time. As a result, there are behavioural modifications that enhance one's physical condition, ego, and integrity. psychological problems, decreased anxiety and sadness, and enhanced social skills. Yoga practices have been linked to positive outcomes including improved functioning, academic success, attention, perceptual sensitivity, reaction time, memory, self-control, empathy, and self-esteem.

The importance of neurofeedback training in cognitive neuroscience cannot be overstated. This study uses electroencephalogram (EEG) data to provide the central nervous system feedback in real time. The brain stays more productive andefficient when subjected to neurofeedback. This approach seeks to improve healthy children's working memory, speed up older people's cognitive processes, and raise their ability for learning.

Alpha, beta, delta, and theta waves are all produced by electroencephalography (EEG) impulses, and each has its unique characteristics. This study focuses on theEEG signals in the Upper Alpha (10–13 Hz), Peak Alpha (10–11 Hz), Beta (13– 30 Hz), and Theta (4–8 Hz) frequency bands.

One of the most observable EEG occurrences of brain waves is the alpha band, which originates from the occipital lobe and occurs during wakeful relaxationwith closed eyes. The alpha wave band is lessened while the eyes are open, duringsleeping, or when you are sleepy. The strongest EEG brain signals are alpha waves, which are generated by the occipital lobes while the eyes are closed.

Alpha-band oscillations, which have an inhibitory effect, are the predominant oscillations in the human brain. Alpha wave brain oscillations occur between 8and 13 Hz. Cognitive processes including as attention, perception, working memory (WM), and long-term memory (LTM) are all impacted by alpha oscillations. Increased Alpha signal has a significant influence on cognitive performance over the whole alpha signal spectrum and is very responsive to demands on semantic memory. Peak PAF, or alpha frequency, is often linked to mental capacity. A decrease in PAF results in physical fatigue. PAF has a negative relationship with age and is attenuated in Alzheimer's patients.

The goal of NFT research is to enhance working memory performance by elevating the higher alpha band of electroencephalography (EEG). Throughoutthe course of five sessions, patients were instructed utilizing visual cues as feedback. EEG signals may be recorded using a single channel EEG system. Eachparticipant's Higher Alpha band was computed using their own alpha frequency band. A single channel electrode was used to capture the signal from the occipital lobe, which was then amplified by an EEG amplifier and connected to the system via a Data Acquisition Device (DAQ).

The signal in the upper alpha band (10–13 hz) was retrieved using lab view software. The patient was then told to unwind by gazing at the depicted natural environment on the gadget. The time it took to edit 10 images during each sessionwas noted. People pay closer attention and concentrate more on the visual presentation as they get more involved in the therapy. High alpha waves thus take priority over low alpha waves. The brain produces alpha waves when a person is relaxed.

As a consequence, when training is complete, the NFT enhances activity by teaching the brain to produce alpha waves on its own. The final results of thisNFT setup showed that the participants were able to learn and that their amplitudelevels in certain alpha bands were growing. The performance of working memory was significantly improved by the rise of the higher alpha band.

Subsequent investigation revealed a positive relationship between individual Higher Alpha band training relative amplitude increases and enhanced capacityfor working memory. Cognitive capacity was tested in the first and last sessions using a conceptual span exam, and the results were compared. All twenty-five individuals in this study demonstrated significant training success.

A common sign of ageing for seniors is a slowing of the intellect, memory, and thinking. The brain may benefit from mental exercise in order to stay sharp. A single channel electrode was used to collect the EEG signal from the occipital lobe, which was then amplified by an EEG amplifier and connected to the system via a Data Acquisition Device (DAQ).

The strongest alpha band (10–11 hz) signal was retrieved using lab view software.The NFT technique was used to provide the neurofeedback training, and as a consequence, the participants' cognitive processing speed increased.

The participant finally learnt to maintain the appropriate brain wave state after having their brain wave frequency measured and analysed during the training. The participant was made to watch animation or listen to music to supply the lab view software with visual clues.

Twenty individuals between the ages of 60 and 65 were chosen for this training. This study set out to determine if the training the older participants got had any impact on their ability to digest information quickly.

For beta/theta training, the high frequency and low frequency components of the recorded EEG were retrieved, together with their power spectral densities. Once the individual fixed his or her attention on the visual screen, the patient was shown the information that had been obtained about brainwave frequencies (/ratio). Changes in the / ratio were reflected in the visual display.

This visual feedback system taught the brain how to operate more steadily and with more control. When the desired brainwave was reached, the brain wasrewarded with visual signals. The unpleasant symptoms were significantly reduced, and by sharpening the subject's concentration, the brain was retrained.

The beta/theta NFT configuration was developed as a consequence to aid healthy volunteers in improving their learning capacity. The NFT procedure is non- invasive and painless. It has enduring effects and may be done again at any moment.

The instruction is carried out using Lab View software. The subject's whole functioning—mental, physical, and emotional—also becomes better. Hence, a user-friendly, low-cost Neuro-feedback setup may be developed to examine the cognitive function of young volunteers and elderly people.

Keywords: - EEG, Cognitive Training , Neuro analysis , Mental Health

TABLE OF CONTENTS

CHAPTER 1
INTRODUCTION

1.1 AIM

- The overall goal of this study is to compile data from a large number of studies to show that YBSR (yoga-based stress reduction) may assist a widespectrum of people manage with clinical and nonclinical concerns.
- The major goal of this research is to investigate the effects of various yoga practices and determine which style of yoga is best for stress reduction among the Millennial Generation.
- Individual alpha Neuro-feedback training was used in this research toenhance working memory performance.
- The experimental findings demonstrated that the participants were able to learn to raise the relative amplitude in specific alpha bands during Neuro- feedback training, and that the amplification of upper alpha greatly enhanced working memory performance.

1.2 OBJECTIVE

The following are some of the research's goals:

- To determine the impact of yoga on stress reduction on humans via mindfulness training.
- To use self-assessment questionnaires to study levels of felt stress, anxiety,depression, academic score, and sleep disturbance among the Millennial Generation.
- To compare the approaches employed by control and normal groups to cope with stress, anxiety, depression and sleep disorders. Using objective

evaluation using electroencephalogram, measure electrophysiological activity during yoga activities, before and after yoga (EEG)

- To determine which varieties impact of yoga are most effective inrelieving anxiety, depression, and stress.
- The Brain Wave Signal Study Software was used to do a stress analysis forhumans.
- To create a Neuro-feedback system that enhances specific aspects of thebrain waves while removing the unwanted ones.

1.3 PREFACE

Living is a type of art. Yoga is a holistic way of life that incorporates all components of ancient Yoga teachings to produce a meditative practice that brings the body, mind, and soul together. In addition to a series of basic yet effective yoga postures and breathing exercises, a greater emphasis is placed on the inner experience of meditation for the well-being of mind and other hidden aspects of human existence. Living in sync with oneself, we believe, makes life easier, happier, and more rewarding. Meditation and Yoga may cause neurological changes and promote physiologic health advantages by improving immune function and lowering arousal levels. The National Institutes of Health has discovered that when meditation is used in conjunction with other treatments,patient results improve.

Effective functioning, memory, academic performance, attention, and self- esteem are some of the additional advantages of contemplative techniques. Yogictechniques were created in ancient times to aid in the realization of the maximum level of functional harmony in the body and mind. Yoga, an ancient Indian discipline, uses intentional breathing control to make breathing regular and to relax the mind.

When compared to a comparable community-home physical activity group, the community-home yoga group showed a higher drop in heart rate, respiratory rate, and an increase in skin resistance at the end of six months. Yoga was shown to help reduce physiological indicators of stress as a result of these changes. The nature of physiological alterations associated with brain functioning that possibly occur with different types of yogic practices is not well understood.

Nowadays, scientists are more interested in yoga, and a huge number of researches are being conducted to better understand the effects of yoga on three parameters: physiological parameters, perceived emotional states, and cognitive performance. Previously, study was restricted due to the fact that

1. They did not address how yoga enhances one's mood.

2. Scientific techniques are lacking

3. There is a lack of evidence on which yoga practices increase visual attention and focus.

This highlights the need for further study on the effect of yoga and its health benefits.

Biomedical signal processing aims to extract useful information from physiological signals collected in the human body. The main goal of biomedical signal processing is to separate the signal of interest from the noisy background and minimize the redundant data stream to a few key parameters.

Electroencephalography (EEG), Electrocardiography (ECG), Magneto encephalography (MEG), and other biomedical signals are examples. Both independent component analysis (ICA) and principal component analysis (PCA) may be used to investigate these signals (PCA). The primary goal of signal analysis is to comprehend human behaviour. Psychopathology, education, genetics, medicine, and physical and developmental disorders are all areas where behaviour analysis is progressing.

There are a variety of ways for analyzing human behaviour and one of the most advanced is physiological signal analysis. The brain signal is utilised in our suggested method to determine the human's cognitive condition. The brain signal is used as the input, which is a mixed signal input in digital signal processing Source separation challenges are used to separate valuable signals from a mixtureof signals of interest, other signals and noise.

The cocktail party dilemma is the most prominent example of a source separation problem, in which a large number of individuals are chatting one after another in a room and a listener is attempting to follow one of the conversations. Using neurons in the auditory area, the human brain can distinguish between the sounds of various persons (source separation).

However, given the current methodologies, source separation utilizing digital signals is a challenging process. Several ways to resolving this problem have been offered, however the majority of the strategies are still in the early phases of development. Principal component analysis (PCA) and independent component analysis (ICA) are two methodologies that have been successfully evaluated.

When there are no delays or echoes, both ways are effective. The blind source separation technique has a broad range of applications, including the analysis of biological signals and photographs. The ICA technique is used to solve the source separation issue in our suggested system. The statistical signal processing method of independent component analysis (ICA) is used to separate mixed sounds, pictures, and signals.

The primary idea behind ICA is to find the underlying independent components in a mixture by seeking for a linear or nonlinear transformation and minimizing statistical dependence. Due to the computational complexity of ICA and commonly used datasets, the ICA technique takes a long time [1].

To decrease the complexity of the algorithms, ICA's VLSI implementation employs modularity, hierarchy, and parallelism. It becomes more efficient as the cost function, which measures component independence, is enhanced. A system level design of ICA employing the evolutionary based optimization approach Shuffled Frog Leap Optimization Algorithm is offered for EEG data processing [2].

The permutation ambiguity is resolved to a considerable degree by using evolutionary computation-based optimizations with extra mutation and crossover processes. This also facilitates the algorithm's convergence to a global optimum.

CHAPTER 2

NERVE CELLS & ELECTRODES

Brain is composed of fundamental units called Neurons. There are about80 billion neurons in the Brain which comes in many forms and shapes. The basic function of the neuron is communicating and transmitting signals to other muscles, glands, cells and vice versa.

2.1 NEURON STRUCTURE

Neurons are classified based on the type of work they are involved into as Sensory Neurons, Motor Neurons and Inter Neurons. The Sensory neurons carry the information from the sense organs, the motor neurons controls the involuntary actions and carry the signals from them and other neurons are Inter neurons. The Neuronal architecture can be divided into 3 main parts as Cell Body, Axon and Dendrite.

Fig 2.1 Structure of Neuron

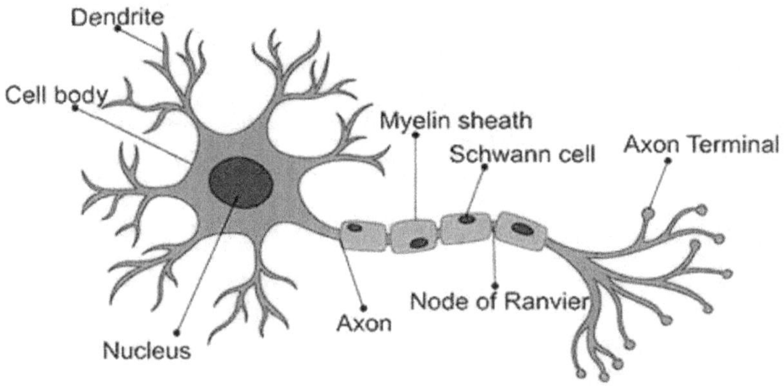

The Cell body otherwise called soma consists of all the necessary components like nucleus, ribosome, endoplasmic reticulum, mitochondria, Golgi bodies and other cellular components. The Axon is a long thin tube like structure which carries information from one dendrite in the top and other dendrite in the end. It is covered by a sheath material called myelin sheath like an insulating material. It has a furrow in intervals called Nodes of Ranvier and Schwann cell isalso present in the myelin sheath which helps in increasing of conduction in the neurons [3].

Dendrites are the nerve endings of the neurons. It's a branch like projections and the main function is to receive message from the dendrites of other neighbouring neurons.

The main function of the neurons is to transmit the signals, either within a single neuron or many neurons form as a chain and transmit the signals. One neuron's axon will connect to another neuron's dendrite through the electrically charged chemical flow from synapse which is present between them. So likewise the chain goes on and on receives and send signals in a fraction of time from various parts of the body to the brain and vice versa [4].

2.2 NEURAL OSCILLATIONS AND ELECTRODES

The brain consists of billions of neurons and these neurons communicate with each other with electrochemical signal which flows as ionic currents through synaptic fibres. And when all these neuronal activities are synchronized gives rise to voltage fluctuations which is known as Neural Oscillations orotherwise called Brain Waves [5].

These neuronal fluctuations have two states known as up state and down state. The upstate denotes the neuronal depolarization along with firing of action potential where as in down state is associated with resting activity and hyperpolarisation. Because of these Neural Oscillations happening in the brain these activities are much stronger that it could be even detected outside the brain and skull.

These electrical activities of the brain could be acquired with the help of electrodes which is placed on the scalp and later it could be analyzed, verified, compared or visualized for further use.

The brain waves occur in repetition between 30 to 100 cycles per minute and these cycles are called frequencies which are measured in Hertz (Hz).

2.2.1 Types of Brain waves

According to the frequency range of the Brain waves they are divided into 5 types as Delta, Theta, Alpha, Beta and Gamma Brain waves. As the frequency of the waveform cycle is measured for 1 second in Hertz (Hz), it consists of positive, negative and neutral vibrations of a cycle [6].

The heights are called amplitudes which determine the strength of the brain waves, so in a waveform we can see both the frequency and amplitude. Many frequencies can occur at the same part of the brain at the same time. These characteristics relate to the type of brain activity which is involved.

Depending upon the situation, emotion or work of the individual the power of each waveband varies having both positive and negative characteristics.
Dr Hans Berger was the first one to record it in humans in 1925 for detecting the brain activity.

2.2.2 EEG Electrodes

The Neuro physiological measurements are acquired with the help of Electroencephalogram (EEG) Electrodes. There are many types of EEG electrodes unlike other bio signal acquiring electrodes based on the attributes, characteristics and convenience of usages. These EEG electrodes are able to sense the synaptic potentials arising from the neurons in the brain either with orwithout any task given to the subjects or patients [7].

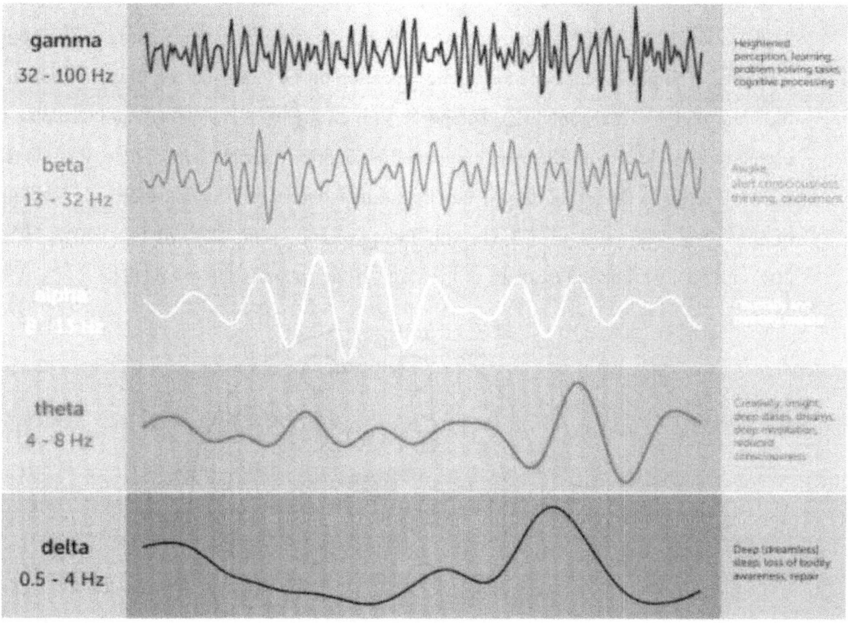

Fig 2.2 Types of Brain Waves

Based on the transmission of the EEG signals from the electrodes to the respective monitors or computer systems there are wired and wireless transmissions [8]. The wired transmission was the one which was used from the startwhich consists of wires arising from each electrode and is directly connected to the monitoring or acquiring system [9]. As the technology in communication rose to another level the EEG electrodes also tend to have wireless transmission through Wi-Fi, Radio Frequency or Zigbee protocol communication. Though wired electrodes are messy and difficult but provide accurate signals compared to the wireless electrodes when compared to the wireless electrodes because of the artifacts [10].

Placement of EEG electrodes on scalp are done with the help of 10/20 Electrode placement system. Those are called EEG Montage which comprises of a set of electrodes being placed in the scalp. The international 10-20 EEG placement system ensures the standard electrode placement spacing which is proportional to the size and shape of the skull [11].

The routine EEG Montage comprises of the 19 channel electrodes along with ground and reference. The placement depends on the location of the electrodes on the lobes which covers the brain region. The Frontal (F), Temporal (T), Central (C), Parietal (P) and Occipital (O) are used as codes along with the even numbers 2, 4, 6, 8 for the right hemisphere, the odd numbers 1, 3, 5, 7 refers to the left hemisphere of the brain and the z refers to the electrodes placed on the midline. The electrodes T3, C3, Cz, C4, T4 are placed on the skull of 10%, 20%, 20%, 20%, 20% and 10% of the total distance of the skull from nasion to the inion [12].

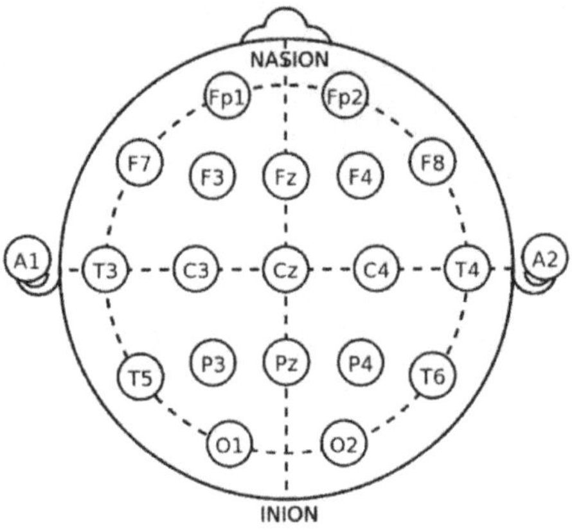

Fig 2.3 10 -20 EEG Electrode Placement System

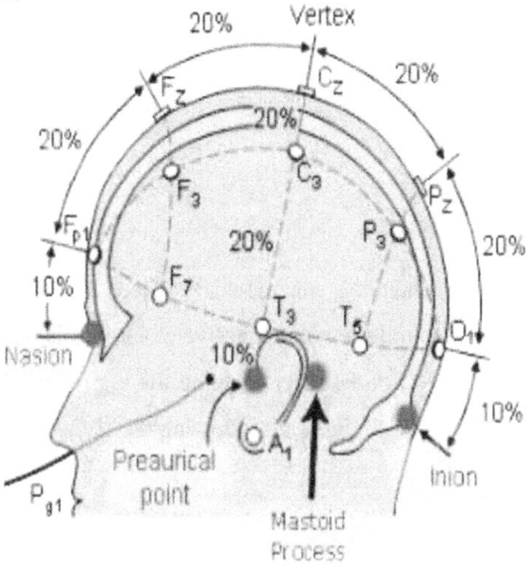

Fig 2.4 Side View of EEG Electrode Placement

Depending upon the needs of the acquirement of the EEG signals and purpose, the EEG electrode channel numbers differ accordingly. In hospitals and other patient monitoring for surgeries, multiple channels maximum of 64 to 256 are used. The electrodes may be made like a cap and worn in some cases [13].

Though multi-channel electrodes are complex to use and to keep in position, but it gives accurate EEG signals [14]. Single or 2 and 3 channels of EEG electrodes are used now a days to get a particular Brain wave either for research or study purpose or for any externally implemented controlled machine to makeit ease to use. These kinds of minimalistic electrodes are very easy to use than the multichannel electrodes and it makes the instrument handy [15].

Based on the placement of electrodes location it is divided into Invasive and Non-Invasive electrode placement. For the people who have epilepsy or seizures who is undergoing surgery the EEG electrodes are placed on the brain inside the skull for measuring the action potential or cortical potential of the brain. The Non-Invasive kind of electrode is placed on the scalp by applying Gelfor adhesion [16].

2.2.3 Location of Electrodes

Depending upon where the electrodes are placed it is divided into two types as Non-Invasive and Invasive electrodes. Non – Invasive electrodes is placed on the scalp of the head and it requires no surgery to do it. After cleaning the scalp EEG Electrode Gel is applied and EEG electrodes are placed on desiredlobe following the 10/20 EEG Electrode Placement [17].

Whereas the Invasive EEG electrode requires surgery then it is placed on the exposed brain directly. This kind of Non – Invasive placement of electrodes are done for Epileptic patients to localize the focus of seizures and for the Parkinson's disease treatment involves combination of this non-Invasivemicroelectrode with deep brain simulation, that is it not only records the EEG signal but also injects electrical current to improve the motor skills for the patients [18].

Even though the Non-Invasive electrode usage is simple and easy to use but as the placement is far from the originating source, so the picked up signals will be containing artifacts. So compared to Invasive electrodes, Non-Invasive electrode signals will be a bit lacking in accuracy but could be minimized by using filters [19].

2.2.4 Number of Electrodes

Number of electrodes used in EEG signal analysis ranges from single electrode to maximum of 256 electrodes. This could be determined for what purpose or task it will be used.

CHAPTER 3

LITERATURE REVIEW

1. **Juris Porozovs[1]*, Viktors Veliks[1] , Aija Kļaviņa[2] , Gunita Praulīte[1](2021)**
Communication technologies enable the use of contemporary and timely efficient analysis of EEG waves, as well as information processing methods that allow for the efficient incorporation of this method intoinstructionally and psychologically oriented research. The goal of this study was to design and validate a way of using EEG signal spectral characteristics in the studies of cognition in the process of music perception via the use of professional studies. The study included 23 research participants, all of whom were University of Latvia students, with participants classified as "non-musicians" or "musicians." The EEG recording was synced with the musical sound by utilizing the createdsynchronization signal, which was fed into one of the EEG device's uni polar input channels. The study examines the underlying rhythm of EEG variations in maximal bandwidth and wave amplitude power in processes related to music perception and cognition at 15-second periods. When artists listen to chorus songs, the range frequency of the range rhythm of alpha and beta does not vary, but when they listen to instrumental music, itgrows, but it is more intense in the band of beta resonance. Non-musicians reacted differently: while listening to chorus songs and instrumental music,the frequency of alpha waves increased while the frequency of beta waves fell.

2. D Calvetti[1] B Johnson[1] A Pascarella[2] F Pitolli[3] E Somersalo[1] B Vantaggi[4] (2021)

Practitioners have long asserted that meditation has a favorable influence on emotion and temperament control, and there has been a concerted attempt in recent years to offer a more exact description of the changes generated by meditation on the human brain. Longitudinal studies have found structural changes in cortical thickness and volume in certain brain areas as a result of meditation practice, which is viewed as proof for its effectiveness beyond subjective self-reporting. Evidence based on real-time monitoring of the meditating brain by functional imaging methods like as MEG or EEG is still difficult to come across. Using Linear Discriminate Analysis (LDA), we show that the data belonging to distinct contemplative or resting states include enough fingerprints of the brain state to allow separation across states, and weidentify the brain areas that appear to contribute to the separation. Our data imply that the cingulate cortex, insular cortex, and several internal structures, including the accumbens, caudate, and putamen nuclei, thalamus, and amygdalae, stand out as separating areas.

3. Sajjad Basharpoor, Fazeleh Heidari & Parviz Molavi (2019)

The frontal lobes govern executive functions, which are high-level cognitive processes that allow for the production of flexible and adaptable goal-directed actions. The goal of this study was to evaluate the association between frontal EEG coherence in theta, alpha, and beta bands and executive functioning in adults. This study includes 168 students as participants. EEG recordings were made at Mohaghegh Ardabili University's psychology laboratory (Iran), and the intra hemispheric and inter hemispheric coherence of frontal areas was examined using the Neuro Guide program. The Adult Executive Skills Questionnaire was given to the participants. Confirmatory factor results established a positive association among executive control and EEG coherence in theta, alpha, and beta bands in frontal regions of the left hemisphere, alpha and beta band coherence in forebrain of the right brain, andalpha band coherence between frontal regions of the two cerebral hemispheres. The regression analysis also demonstrated that synchronization of the alpha, beta, and theta bands between the left and right frontal areas, as well as consistency of the beta and theta bands in the left frontal regions, indicate mental activities. These findings suggest that frontal cortex activity, particularly in the left hemisphere, is linked to executive functions and cognitive processing.

4. **N N Abd Razak[1] , K A Sidek[1]* , K Y D Alaa[2] and N F Mohd Nasir[3]** Depressive moods are becoming more common in today's culture as aresult of people's hectic, competitive, and demanding lifestyles. These disorders had grown quite frequent, particularly among young and middle-aged people, and the World Health Organization had listed suicidal thoughts as one of the major causes of mortality (WHO). In neurological study, nature sound (sound of rain or the seashore) has been connected to depression and anxiety, and it has been found to be an alternative to anxiety relief. The electroencephalographic (EEG) waveforms has beenshown to be capable of recognizing information from brain signals as wellas data from the past via Bluetooth connectivity. A few tests resulted in thewaveforms used in the study. As a result, the study suggests that the association between the reaction of EEG signal on brain relaxation withother types of portable unit be investigated in order to help perceived stresspatients attain a relaxed or tranquil state.

5. **ANthony d. Bateson [1] and Aziz u. R. Asghar [2] (2021)**

The study's goal was to design, build, and test a general-purpose EEG platform that connects with a Smartphone. The goal was to create a system with 19 EEG channels and data saved on a Smartphone through a Wi-Fi connection. The hardware was created using three ADS1299 integrated circuits, and the Smartphone software was created using the Unity game engine. The device was evaluated by capturing alpha waves in subjects duringtimes of eye closure (BlandAltman' statistical comparison with a clinical grade EEG system). The smartphone was also utilized to transmit time-locked audiostimuli using an unusual paradigm to assess the created system's capacity to gather event related potentials (ERP) while sitting and walking. There were no significant changes in alpha wave peak point, wavelength, or area under the curve when comparing intra-system (two consecutive periods of alpha waves) or inter-system (built smartphone-based EEG system vs FDA- approved system). The maximum amplitude of the auditory P300 response to aberrant tones was considerably larger as compared to standard tones for sitting and walking activities, according to ERP data. Our broad sense EEG system is intended to inspire other researchers to create and construct their own customized variations instead of being constrained by the fixed features of commercial devices.

6. **Andrés-Eduardo Nieto-Vallejo; Omar-Fernando Ramírez-Pérez; LuisEduardo Ballesteros-Arroyave; Angela Aragón (2021)**

   Yoga is a type of mental exercise with emotional and intellectual advantages that may improve attention, mental health, and regeneration. However, the learning process is difficult since meditators do not receive instant feedback indicating whether or not they are performing the task correctly. EEG Neuro feedback training is one way for teaching the brain to regulate itself, and it has the potential to improve the effectiveness of meditation. The advantages, however, vary substantially amongst people with a large percentage of lack of effectiveness. An EEG Neuro modulation Tutorial Mode users ' self design approach is presented in this study to deliver real-time performance feedback to mediators to raise levels of concentration and relax via a visual, auditory, and scent stimuli interface. While meditation practice, nine individuals' levels of concentration and relaxation were assessed using a mobile Neuro sky EEG headset biosensor to analyze the occurrence of each sort of stimulus duringactivity. When compared to a meditation session without any stimuli,visual stimulus feedback was able to enhance the concentration power of78 percent of the participants by 11.8 percent.

7. **Yu-tzu lin 1,2, (member, ieee), Yi-zhi liao1,2, xiao hu 3 , and Cheng- chih wu1,2(2021)**

The purpose of this research is to investigate cognition throughout program interpretation using physiological evidence by recording and comparing electroencephalogram (EEG) activity in different bandwidths and eye movements of individuals with high or low programming talents. Thirty-three undergraduate students majoring in computer science took part in the study. We measured their EEG activity while they read two programs with three different sorts of program structures. At the same time, an eye tracker was used to capture the participants' eye movements inorder to better understand the link between the program understanding process and EEG activity. The innovative studies indicate that high participants performed better in program execution of tasks of next several for memory recall (theta power), attention resource allocation (lower alpha power), and interaction between working memory and semantic memory (upper alpha power), proving related theories proposed in previous research on programming and cognition. The findings of this study not only give objective proof of the functions cognition plays in program understanding, but also offer educators advice for developing appropriate instructional tactics.

CHAPTER 4

YOGA – ART OF LIVING

Yoga is to revive, build, and discover your real self with these exercises and poses. Living is an art. Yoga is a whole system for recharging and uniting your mind, body, and soul. This ancient wisdom adapted for current times may simply be incorporated into your daily routine, resulting in a lasting sense of inner strength and total well-being.

For more than 5000 years the Yoga form of art is prevailing in India. Though many people think of yoga as only a physical activity in which individuals twist, turn, stretch, and breathe in intricate ways, this is simply the most basic component of this deep discipline of awakening the unlimited potentials of the human mind and spirit. Yoga is a discipline that incorporates thetruest expression of living.

Yoga is said to be around 10,000 years old. The first reference of the custom may be found in Nasadiya Sukta of the Rig Veda, the Ancient literature. Its origins may be traced back to the Indus Valley Civilization. A figure sitting ina yoga stance appears on the Pashupati seal from the same civilisation, proving its ubiquity in ancient times. However, the oldest Upanishad, Brihadaranyaka, has the first description of the practices that subsequently formed part of yoga. Pranayama is mentioned in one of its hymns, while Pratyahara is mentioned in the Chandogya Upanishad.

4.1 TYPES OF YOGA

The following are the various types of Yoga which was followed by the subjects pertaining to this research which was made to relieve the stress in this modern day world.

Balasana - Child Pose

The most peaceful yoga posture for calming the mind and renewing the body is Balasana, or child position. It's also a fantastic yoga pose for easing lower back discomfort, hip strain and stiffness in the shoulders and neck. It has a relaxing and soothing effect that is beneficial to both mental and emotional well- being. Kneel on your yoga mat with your knees pulled together and yourbuttocks resting on your feet. Exhale as you drop your body over your thighs, keeping your head on the mat and your hand on the floor. Return to the beginning position after 30 seconds to 1 minute of holding the pose and breathing appropriately. Following that is the first of two repetitions. Rep 3–4 times more.

Fig 1 Balasana - Child Pose

Uttanasana - Forward Bend Standing

In yoga, the standing forward bend is an excellent approach to release tension. It improves blood flow throughout the body, allowing the neurological system to function appropriately. It also helps with spine stretching, relaxation, and enhanced flexibility. Because it tones the abdominal muscles and organs, it isan excellent weight-loss yoga position. Starting by standing with your feet together and your arms to the side is a nice place to start. Inhale by raising your hand over your head, then exhale by bending your body forward. Maintain a straight back and both hands on the ground. Maintain a regular breathing pattern for 30 seconds while holding the stance. Return to your starting point. The firstof two repetitions comes next. This should be repeated five to 10 times.

Fig 2 Uttanasana - Forward Bend Standing

Adho Mukha Savasana – Downward Facing Dog

Down dog, also known as downward facing dog, is a fantastic stress- relieving yoga position that helps to awaken the senses while also reducing tension and tiredness. It helps to enhance bone density and prevent bone illnesses like osteoporosis in women. It energizes the body by increasing blood circulation and may aid in the treatment of neck pain by decreasing upper-body stiffness. While kneeling on your hands and knees, keep your back straight and parallel to the floor. Lift your hips by straightening your legs and elbows and forming an inverted V with your body. Maintain the position for 30 to 1 minute while breathing normally. Exhale gradually and gently return to the starting position. The first of two repeats follows. It should be done five to six times.

Fig 3 Adho Mukha Savasana – Downward Facing Dog

Setu Bandha Sarvangasana - Bridge Pose

The bridge pose is the ideal yoga posture for anxiousness. It's an excellent approach to tone and stretch your thighs and hips while also toning your back and leg muscles. It also aids in the treatment of high blood pressure as well as therelief of backaches, headaches, sleep issues, and fatigue. Lie down on your yoga mat with your knees bent and your feet near to your buttocks. The distance between your feet should be the same as the width of your hips. You may elevate your hips and breastbone by pulling your arms towards the floor and maintaining your head fixed on the ceiling. Return to the beginning position after 30 seconds to 1 minute in this position. The first of two repetitions comes next. It must be done 5–6 times.

Fig 4. Setu Bandha Sarvangasana - Bridge Pose

Savasana - Pose of the Dead

In yoga, the corpse posture, also known as Savasana, is the ideal way to relax. The corpse position, the most soothing and restful of all postures, is a must-do for every yoga practice. It induces a profound state of rest and relaxationin the body. This yoga posture promotes sleep, lowers blood pressure and relaxes the nervous system. Place yourself on the yoga mat and use a cushion or pillow for support. To relax your knees and toes, keep your feet slightly apart. Place your arms on each side of your body, palms facing up. To calm your body and mind, close your eyes and take slow, deep breaths. Stay in this posture for 10 minutes to complete your yoga practice feeling energized and serene.

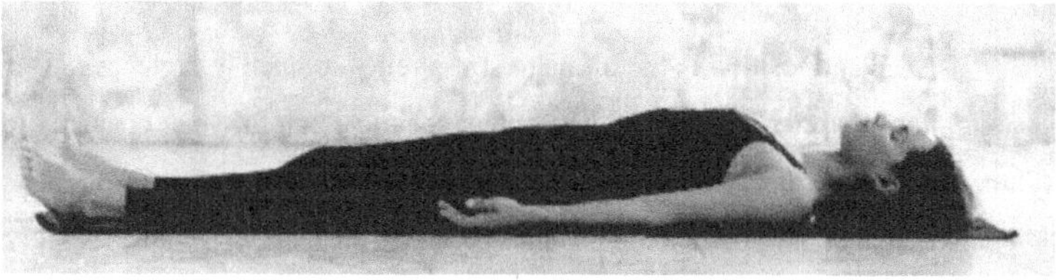

Fig 5. Savasana - Pose of the Dead

Uttana Shishosana - Puppy Pose

The extended puppy stance, which stretches the spine and shoulders, is oneof the most fundamental yoga poses for relaxing the body and eliminating stiffness. It may be used to improve posture, flexibility and mind-body regeneration. Because it tones the hips and back, this is one of the finest yoga postures for postpartum weight loss. Starting on your hands and knees, drop your chest to the ground gently by pushing your arms forward while keeping your elbows off the ground. Drop your head to the floor until it's absolutely flat. Keep your neck relaxed and hold the stance for 30 to 1 minute while breathing normally. Now exhale and return to your starting position. The first of two repetitions comes next. It should be repeated 5–6 times.

Fig 6. Uttana Shishosana - Puppy Pose

Marjaryasana - Pose of a Cat

The cat position is the ultimate stress reliever since it stretches the spine and tones the abdominal muscles and organs. It is one of the most helpful yoga positions for PMS relief and improves overall health by minimizing digestive issues. Get down on your hands and knees, with your wrists behind your shoulders and your knees beneath your hips. Exhale now, curling your spine downward and aiming to the ceiling. Bend your head to the floor, but don't let your chin contact your breast. Return to the starting position after 30 seconds of keeping the stance. The cat position is usually coupled with the cow stance, which is accomplished by inhaling and lowering your belly to the mat while pointing your head upward. Repeat these two stances 5–6 times more, holding each for 30 seconds each time.

Fig 7. Marjaryasana - Pose of a Cat

4.2 IMPORTANCE OF YOGA

Yoga practice aids in the management of an individual's mind, body and spirit. It combines physical and mental disciplines to produce a tranquil body andmind; it aids in the management of stress and anxiety and keeps you relaxed. It also aids in the improvement of flexibility, muscular strength, and body tone. It boosts your energy, vitality and respiration. Yoga may appear to be only stretching, but it can do much more for your body in terms of how you feel, look,and move.

Yoga asana build strength, suppleness and self-esteem. Yoga practice on a regular basis can help you lose weight, decrease stress, boost your immunity, andlive a better lifestyle.

Yoga and meditation are two parts of a comprehensive self-healing strategy that promotes body-mind-soul harmony and aids in the treatment of a variety of diseases.

Self-healing and blood pressure regulation are aided by this supplement. It aids in the slowing of the aging process. Enhances the functioning of the nervous system Improves concentration and attention. Aids in weight loss Losing weight improves flexibility and posture while also increasing energy levels.

CHAPTER 5

EEG ACQUISITION

5.1 RMS MAXIMUS 24 ELECTROENCEPAHALOGRAPH

RMS Maximus is a premium EEG Machine for Neurological Diagnostics. It has 24 channels which draws the full band simultaneous acquisition of rawdata [20]. Data transmission is through high speed USB 2.0. Simultaneously Dataacquisition and Data analysis for the same patient could be done with the software's RMS Acquisition and RMS Analysis respectively. The main advantage of this RMS Maximus is it is compact and light weight and its artifact free recording in any recording. It is compatible with PC and Laptop and Event marking with review and printing options are available [21].

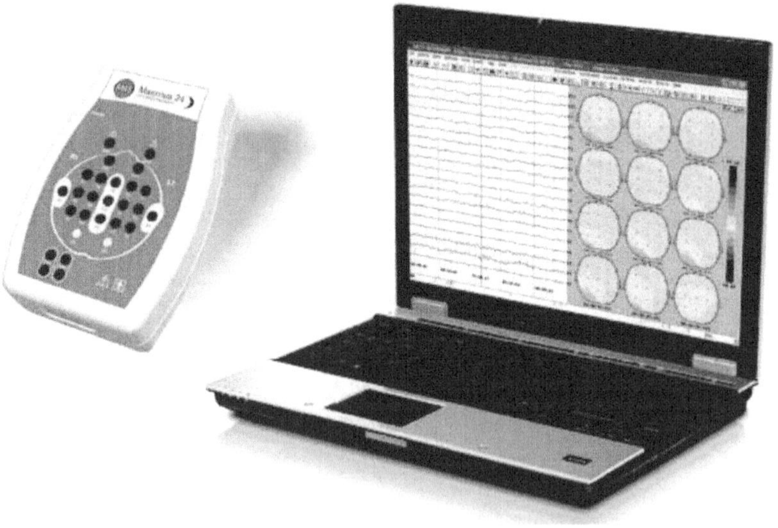

Fig 5.1 Rms Maximus 24 Electroencephalograph

CHAPTER 6

MATERIALS AND METHODS

6.1 PSYTASK

PSYTASK is software that allows you to plan and conduct psychophysics and psychophysiology experiments.

1. Presentation of visual (on the monitor screen) and audio (through sound card and speakers) inputs.

2. Measurement of reaction time. The keyboard and mouse may be used to assessresponse time.

3. PSYTASK integrates with the "Mitsar" digital EEG recorder (Win-EEG software) to provide synchronized stimuli and EEG capture. Two PCs are linked using COM ports in this scenario. On the first computer, the PSYTASK application is used, while on the second computer, the Win-EEG program isused [22]. The Win-EEG application sends synchronization codes to the PSYTASK program, which allows it to regulate its operation and display synchronized stimuli with EEG acquisition. Event related potentials (ERP) and/or event relatedde synchronization may be calculated and analyzed using the recoded data (ERD) [23].

4. The PSYTASK application may use the COM port to deliver synchronization codes to another program. This capability allows you to conduct ERP investigations with different amplifiers and a different application that can identify synchronization codes [24].

5. The PSYTASK application has a user friendly interface that allows you to create psychological activities.

6. The PSYTASK application calculates response time and task performance metrics (omissions and commission mistakes) and saves them in a built-in database for further analysis and report production. Graphical Task for GoNogo.

For 100 milliseconds, one of two highly distinguishable visual stimuli, a coloured square with a tiny square close to either the top or bottom border is shown. Stimuli are provided in pairs, one for each trial. The inter-stimulus interval in a pair is 1000 milliseconds. Only two separate pairings of stimuli are shown: top-to-top ("Go") and top-to-bottom ("Nogo") trials. Trials are provided in a 50 percent chance random order. The period between trials is 1500 milliseconds. In the event of Top-Top pairs of stimuli, the subject must click a button (Right Arrow Key) as fast as possible and disregard Top-Bottom pairs of stimuli. There are a total of 450 trials in this study. For both the Go and No-go conditions, performance parameters are determined.
The research was done on three separate groups: College students, Professional Workers, and Elders.

Age	College students		Professional Workers		Elders	
	Male	Female	Male	Female	Male	Female
17-21	70	70				
21-30			50	50		
30-50			50	50		
51-75			30	30	25	25

Table 6.1 Trial Participant's Categories

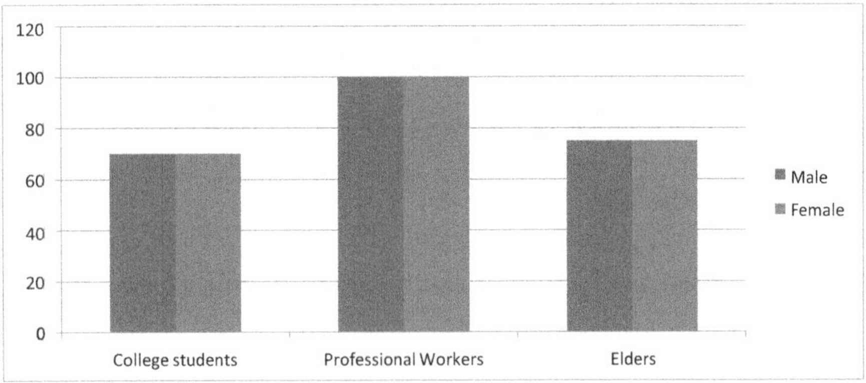

Fig 6.2 Number of participants – Age, Gender, Profession

6.2 DATA ANALYSIS

The basic goal of biomedical signal processing is to separate the signal of interest from the noisy background and minimize the redundant data stream to a few, but important, characteristics.

Both Independent Component Analysis (ICA) and Principal Component Analysis (PCA) may be used to analyze these signals (PCA). The primary goal ofsignal analysis is to anticipate human behaviour. Psychopathology, education, genetics, medicine, and physical and developmental disorders are all areas where behaviour analysis is progressing. There are various ways for analyzing human behaviour, and one of the better ones is physiological signal analysis. The brain signal is utilised in our suggested system to identify the human's cognitive condition. The brain signal is used as the input, which is a mixed signal input in this case. It is necessary for the source separation algorithm to forecast the characteristics of each signal.

6.2.1 PRINCIPAL COMPONENT ANALYSIS (PCA)

PCA is an analytical method that uses an orthogonal transformation to turna collection of observations of possibly correlated variables into a set of values oflinearly uncorrelated variables known as principle components [25]. The number of fundamental components is less than or equal to the number of originalvariables.This transformation is designed such that the first principal component has the most variation possible, and each successive component has the least variance possible while staying orthogonal to the preceding components [26]. The primary components are orthogonal because the eigenvectors of the symmetric covariance matrix are orthogonal. PCA is influenced by the relative scale of the source variables [27].

6.2.2 INDEPENDENT COMPONENT ANALYSIS (ICA)

Clinical sensors or equipment are used to examine biomedical signals such as the electroencephalogram (EEG), magneto encephalogram (MEG), and electrocardiogram (ECG) (ECG) [28]. However, aircraft noise and other unexplained noise signals, such as eye movements, muscle noise, and instrument power noise, distort observed data. The independent component analysis (ICA) technique, which differentiates aircrafts from observable waveforms, might be used to tackle this challenge [29].

6.2.3 FAST ICA

To maximize non-Gaussianity, a statistical independence metric, the approach employs a fixed-point iteration strategy. It can also be achieved via an approximative Newton iteration. To estimate n different components, run these methods n times [30]. We simply need to add a basic orthogonal zing projection to theloop to ensure that we estimate a new independent component each time.

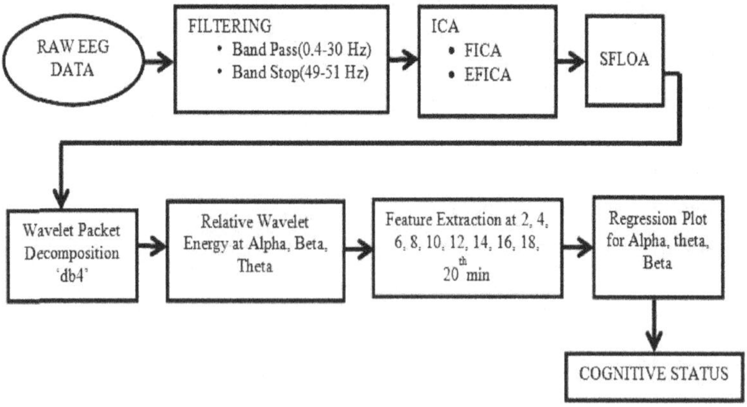

Fig 6.2 Algorithm Flowchart

The Fast ICA technique and its underlying contrast functions offer significant benefits over regular ICA approaches. Convergence occurs quickly. The algorithm is straightforward to put into action. Boosts speed and takes up minimal memory.

6.3 LAB VIEW - PROGRAMMING SOFTWARE

Lab View (Laboratory Virtual Instrumentation Engineering Workbench) is a framework and production environment for a visual programming language developed by National Instruments. The purpose of this type of programming is to automate the operation of laboratory processing and measuring equipment. On a number of platforms, Lab VIEW is widely used for data collecting, instrument control, and industrial automation. The Lab VIEW programming language is a dataflow programming language. The structure of a graphical block diagram (the LV-source code) determines execution, on which the programmer joinsindependent function-nodes by drawing wires.

These wires carry variables, and any node can execute as soon as it has access to all of its input data. Virtual instruments are Lab VIEW programs/subroutines3 (Vis). A block diagram, a front panel, and a connecting panel are all included with each VI. The latter is used to represent the VI in blockdiagrams of other VIs, also known as VIs. An operator can enter data into or retrieve data from a running virtual instrument by using the knobs and indicators on the front panel. A lab view application is used to capture the EEG signal, which is then processed. The relevant frequencies were retrieved using appropriate filters, and then subjected to spectral measurements, integrated, and the alpha ratio determined.

CHAPTER 7
RESULTS

7.1 EEG SIGNAL PROCESSING

Wavelet analysis is a data segmentation technique that divides data into numerous frequency bands. When using DWT to analyze data, selecting the appropriate wavelet and number of decomposition stages is crucial. The number of breakdown levels is determined by the signal's main frequency components. The values are selected in such a way that the wavelet coefficients preserve the signal components that are best suited to the requisite frequencies for signal classification. Because EEGs provide little relevant information above 30 Hz, thenumber of breakdown levels in this study was limited to six. Experiments are frequently carried out using a range of wavelets, with the most successful one selected for the specific purpose. The smoothing characteristic of the Daubechies wavelet of order 4 (db4) was demonstrated in this work to be adequate for identifying changes in EEG data. Table 1 shows the frequency bands that correspond to the six decomposition phases.

Table 2: WAVELET

DECOMPOSITION LEVEL	RANGE OF FREQUENCY BAND
D1	125-250 Hz
D2	64-125 Hz
D3	32-64 Hz
D4	16-32 Hz
D5	8-16 Hz
D6	4-8 Hz

Decomposition Level

7.2 POST YOGA PHYSIOLOGICAL TASK

Subjects who do not practice Yoga are given instruction to practice Yoga on a regular basis for around 6 months. After the time limit has expired, they are given the Go No-go PSYTASK exam again, and their results are taken. Everyone's EEG is taken for the second time. The EEG recorded before and after yoga practice was compared. And it is plainly shown that, according to the statistics, the subject's score is higher than the previous one, indicating that the person's focus has improved. The EEG comparison also shows that, as a result of Yoga, the alpha waves have increased significantly in comparison to the priorone.

Although the students improved in certain areas as a consequence of the yoga intervention, the overall outcomes were minor. These results might be dueto a variety of circumstances. Student engagement was at an all-time high of 80%before yoga was introduced. Student involvement increased by 88 percent after the yoga intervention. This 8% rise isn't substantial enough to be statistically significant.

A Go No-go test sample comparison is presented below, which was collected before and after school pupils practiced Yoga. EEG analysis and acquisition software, as well as an RMS maximums machine, are used to determine the individuals' emotional states. The subject's emotional state is categorized using a tomography map. As input, the raw EEG data is provided. The filtering procedure is done in two circumstances to minimize the noise in the signal. The filtered result is mixed signal data, and the independent component analysis technique is employed to separate the mixed signal. There are various ICA algorithms for separation, thus the convergence time of the signals based on the ICA algorithm is determined to forecast the superior method. FASTICA has afaster convergence time based on the total analysis, hence it was chosen for the separation procedure. Only 7 channels (Fp1, Fp2, Fz, Cz, P3, P4, Pz) were chosen to examine the cognitive state out of a total of 19 channels.

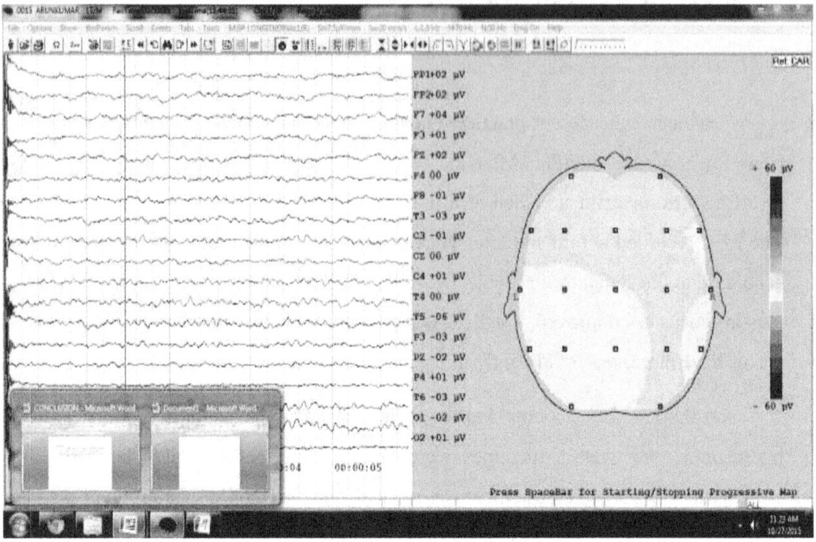

Fig 7.1 Tomography of Brain

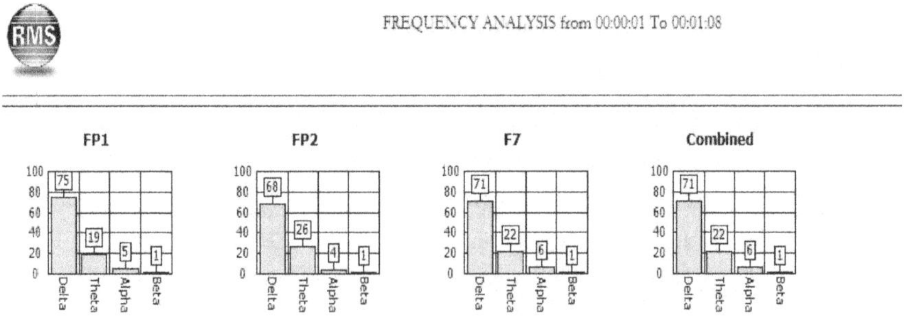

Fig 7.2 Frequency Analysis showing the EEG waves before Yoga

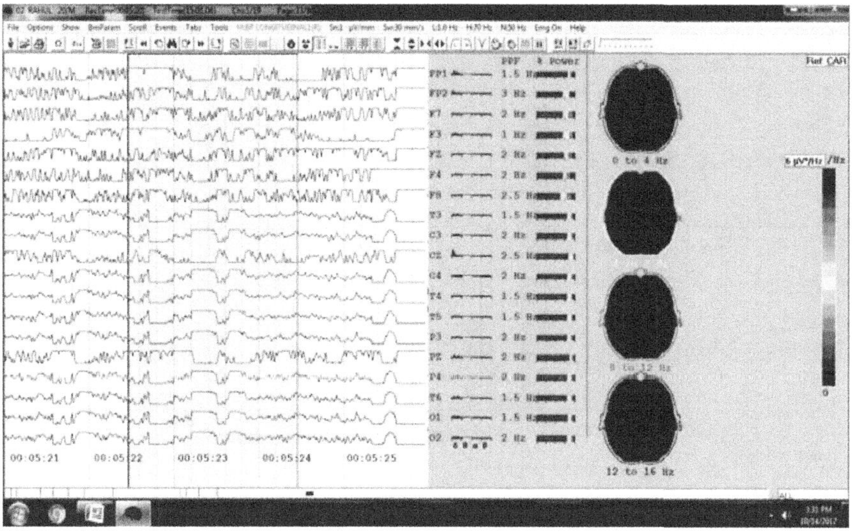

Fig 7.3 Topographical analysis of the post yoga

Attention, executive control, and working memory are among cognitive processes that theta and alpha rhythms are thought to be important indications. Based on theoretical analyses of EEG dynamics and extensive experimentalevidence, short and long range coherence of alpha and theta sources is projected to increase with cognitive activity (Nunez and Srinivasan, 2005). Mental attention on visual tasks was linked to blink inhibition and frontal theta band activity. Figure 6.4 shows that theta band frequency power values (slope) were substantially higher (P0.05) before yoga than after yoga in the Pz, Fz, and CZ electrode locations.

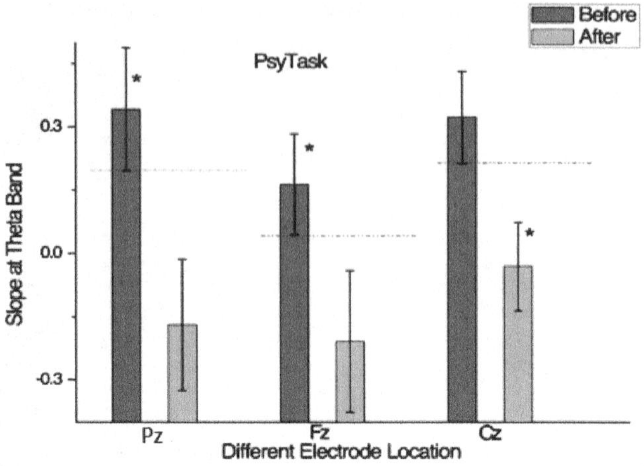

Fig 7.4 Mean and one-standard error for slope at theta band of EEGsignal during Psytask in before and after yoga

Alpha activity implies a reduction in attentiveness, whereas theta activity indicates a stress reaction. Alpha suppression is linked to sensor motor information processing modulation and is more strongly suppressed during the processing of complex tasks requiring perceptual, judgment, memory, and motor demands. For partial lobes of the brain, the slope of relative wavelet packet energy at alpha band was substantially (p0.05) higher before yoga than after yoga.

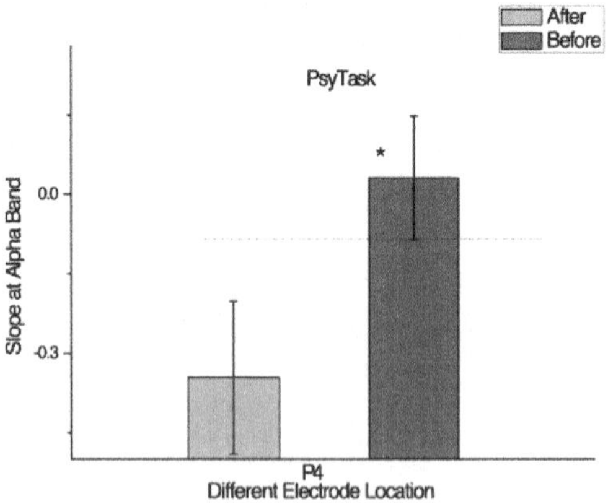

Fig 7.5 Mean and one-standard error for slope at alpha band of EEGsignal during Psy-task in before and after yoga.

7.3 SIMULATION ANALYSIS

7.3.1 PREPROCESSING OF RAW EEG DATA

As input, the raw EEG data is provided; the filtering procedure is done in two circumstances to minimize the noise in the signal. This is done with the Lab View Software. The difference between the raw data and the filtered data is seen in figure.

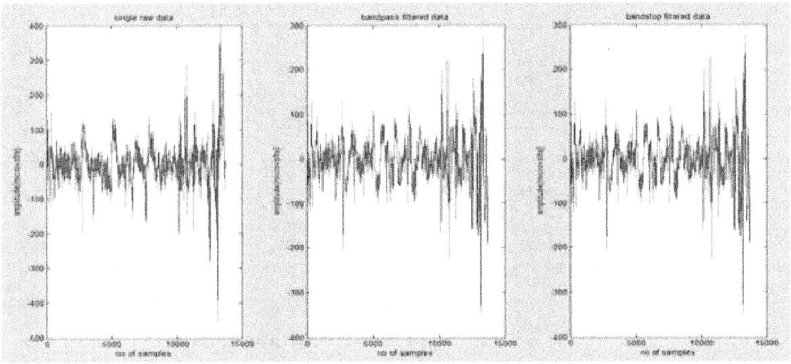

Fig 7.5 Filtered and Raw Signals

The filtered result is mixed signal data, and the independent component analysis technique is employed to separate the mixed signal. The filtered result is mixed signal data, and the independent component analysis technique is employed to separate the mixed signal.

The filtered result is mixed signal data, and the independent component analysis technique is employed to separate the mixed signal. The wavelet coefficients of the signals fp1, fp2, fz, cz, p3 and p4 may be used to calculate E1, E2, •••, E6. For various frequency bands, relative wavelet energy (RWE) 1, 2,

•••, 6 may be determined. The slope value may be used to assess the cognitive condition of individuals. The output of relative wavelet energy is shown in figure

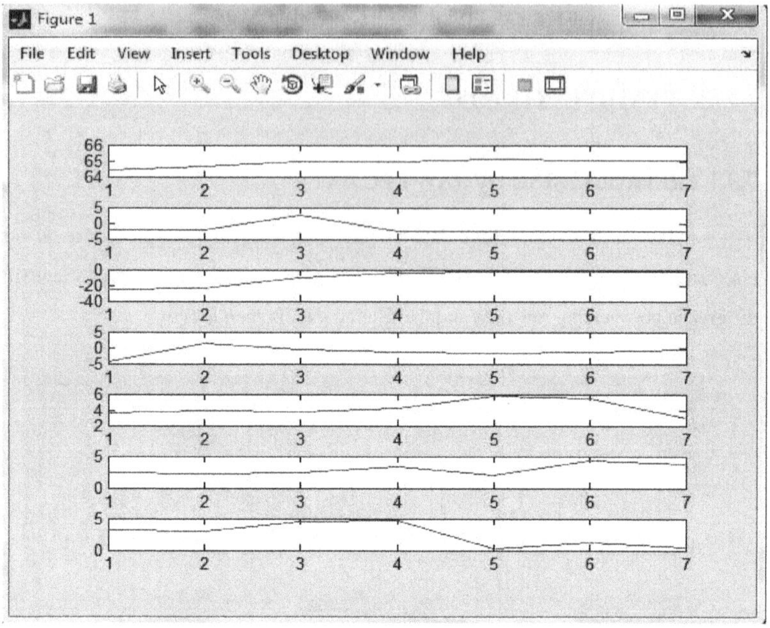

Fig 7.6 Filtered and Raw Signals

7.4 TRAINING RESULTS

The individual begins Neuro-feedback training in order to increase working memory. The training's goal is to increase working memory performance as well as relaxation. As the participants get more trained and thelength of the sessions decreases, the threshold is progressively raised for subsequent sessions.

The training is conducted in a quiet setting with few distractions for theuser. Before the training begins, the patient is forced to rest.

The individuals' scalp electrodes are attached, and the higher alpha is measured for a period of time. The individual is then invited to relax by lookingat natural scenery. For each session, the time it takes to alter 10 photos is recorded.

The subject is pushed to finish sessions quicker as the threshold is raised at eachstep. Initially, feedback was offered for the whole alpha frequency (8-13Hz), as well as the length of the session and the alpha wave amplitude (peak to peak).

The upper alpha (10-13Hz) feedback was then supplied, and the length of the sessions as well as the amplitude of upper alpha were recorded. The followingare the subjects' training results.

Training results for alpha training

NO OF SESSIONS	THRESHOLD	DURATION(Sec)	AMPLITUDE (PEAK TO PEAK)μV
1	0.5	1020	0.0021
2	0.5	990	0.0172
3	0.5	834	0.0623
4	0.7	420	0.0942
5	0.7	288	0.0987

Table 3 Slope Values of Alpha Band

NO OF SESSIONS	THRESHOLD	DURATION	AMPLITUDE (PEAK TO PEAK)μV
1	0.5	1380	0.07142
2	0.5	1080	0.04081
3	0.5	936	0.0912
4	0.7	870	0.1081
5	0.7	680	0.1389

Table 4 Slope Values of Theta Band

7.5 EVALUATION OF RESULTS

The findings are examined by producing graphs in order to see whether theperson has had any favourable outcomes. The outcomes are determined by an individual's performance throughout training sessions.

7.5.1 ALPHA TRAINING RESULTS

Graphs were used to assess the outcome. The length of the Neuro-feedback training session lowers as the session progresses, as seen in the graph below. In subsequent sessions, the amplitude of the alpha was increased.

EVALUATION OF SUBJECT 1:

Name: Preethi G

Age: 21 Training:

NFB Drugs used:

Nil

No. of sessions attended: 5

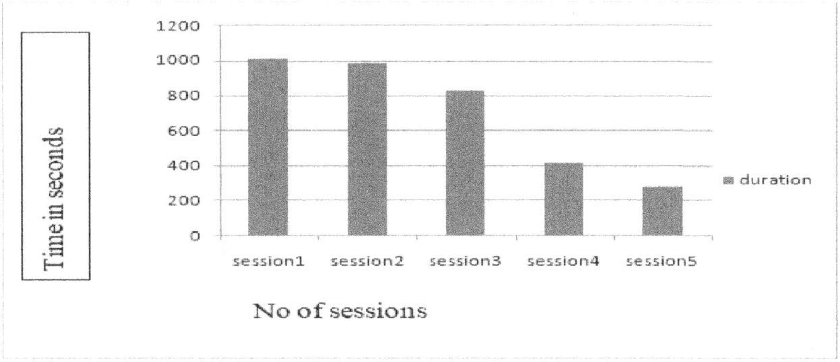

Fig 7.7 Duration of subject 1

EVALUATION OF SUBJECT 2

Name: Aarthika A

Age: 21

Training: NFB

Drugs used: Nil

No. of sessions attended: 5

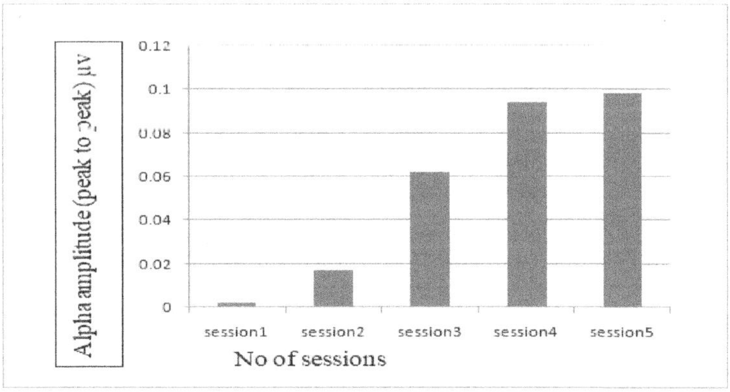

Fig 7.8 Duration of subject 1

7.5.2 UPPER ALPHA TRAINING RESULTS

Graphs were used to assess the outcome. The length of the Neuro-feedbacktraining session lowers as the session progresses, as seen in the graph below. In subsequent sessions, the amplitude of the higher alpha has been raised.

EVALUATION OF SUBJECT 1:

Name: Dhanalakshmi B

Age: 21

Training: NFB

Drugs used: Nil

No. of sessions attended: 5

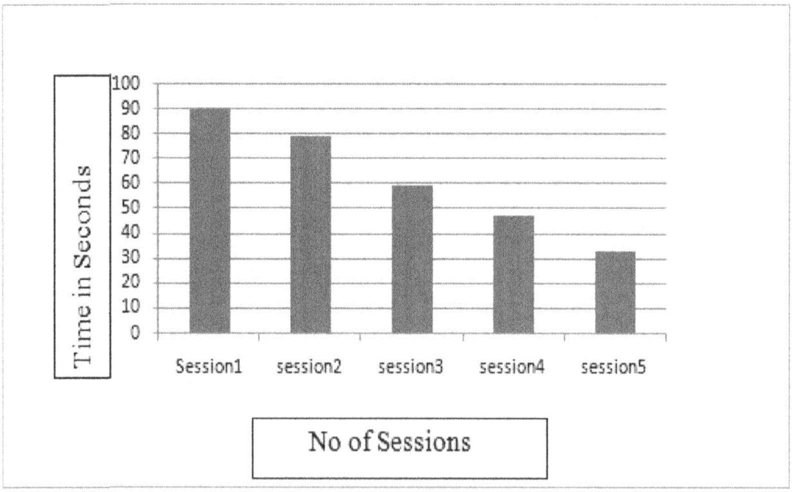

Fig 7.9Duration of subject 1

EVALUATION OF SUBJECT 2

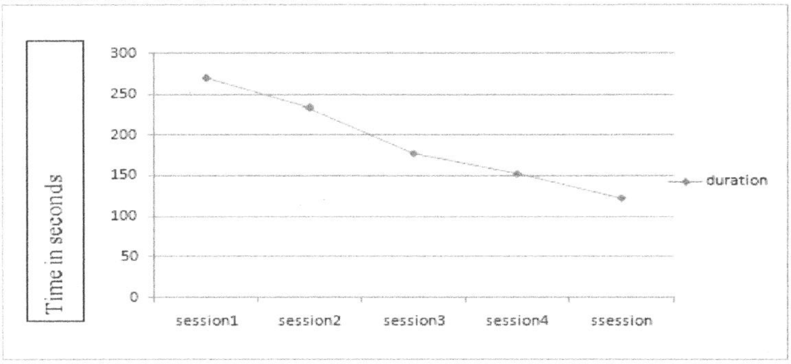

Name: Soma Soujanya

Age: 21

Training: NFB

Drugs used: Nil

No. of sessions attended: 5

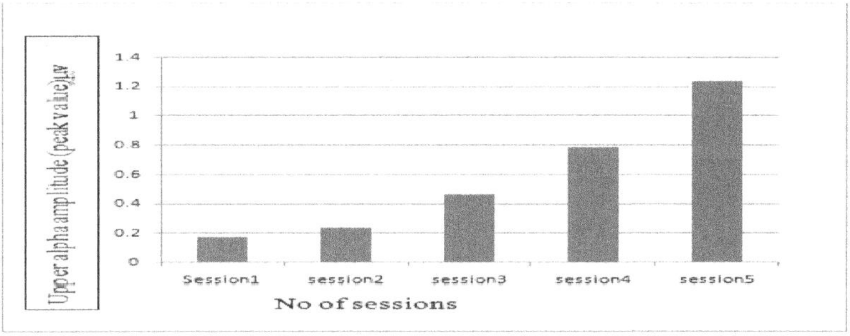

Fig 7.10 Duration of subject 2

7.5.3 COMPARISON OF ALPHA AND UPPER ALPHA RESULTS

The alpha and higher alpha were used to teach the individual, and the outcomes were compared. The time it takes for the individual to relax in alpha training (8-13Hz) is longer than in upper alpha training (10-13Hz), and the upper alpha amplitude values are higher than the alpha. In comparison to alpha training,higher alpha training is an efficient training since it demonstrates considerable development.

CUMULATIVE RESULTS

Fig 7.11 Duration of subjects

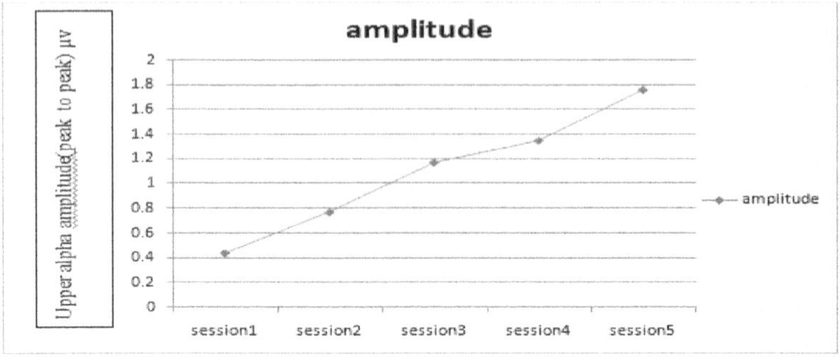

Fig 7.12 Upper Alpha Amplitude of Subjects

Figure 7.11 indicates that as the training progresses, the amount of time spent relaxing decreases. The amplitude of upper alpha has grown significantly as the session progresses, as seen in fig 7.12.

CHAPTER 8
DISCUSSION

Stress has been shown to improve a subject's performance greatly. Though modest amounts of stress might help you stay focused, too much stress can affect your job performance, cause mishaps in the military, cause you to lose sleep and prevent you from changing your routine. One of the most important causes linked to plane crashes is cognitive tiredness. Because physiological signals may be gathered constantly and without interfering with a person's work performance,they are valuable for supplying current subject states. EEG has been shown to bea valid tiredness indicator that may also be utilized as a fatigue countermeasure. The EEG signal's relative wavelet packet energy was utilized to evaluate mental weariness, which impacts cognitive processes and plays a role in the loss of attention.

The research by Belyavin and Wright (1987) discovered a link between EEG and alertness. Sharp reductions in beta activity are a helpful strategy for lowering alertness, as shown by the findings of this research. Alpha waves are associated with relaxed states, lowered attention levels, and a sleepy yet awake state, whereas beta waves are associated with alertness levels. The performance of a vigilance task improves as the beta band activity rises. Alpha activity impliesa reduction in attentiveness, whereas theta activity indicates a stress reaction. For example, following yoga, alpha activity in the frontal site decreased, but beta activity in the central, frontal, and parietal sites increased dramatically. The findings show a significant improvement in the individuals' attentiveness.

Fast waves are connected with enhanced alertness, arousal, and excitement, and are sometimes referred to as beta rhythms. The relative wavelet packet energy of beta beats rises considerably following yoga practice. Significant variations were detected in the frontal sites for alpha activities, as well as the frontal and parietal sites for beta activities, according to the findings of the present research.

Previous research has shown that when attentiveness falls, the EEG spectrum shifts. Low frequency EEG waves such as theta and alpha rhythms, forexample, may rise in proportion while higher frequency waves such as betarhythms decrease (Gevins et al., 1990). Beta rhythms drop before the yoga task in the current research, indicating that respondents' attentiveness levels fall whencompared to when they are not doing yoga. In comparison to not doing yogabefore a cognitive strain, our findings showed that doing yoga before a cognitive task boosts brain activity adequately. Concentration, cognition, and listening areall connected with beta waves. In the absence of muscular action, synchronizedbrain waves represent attention, awareness, or consciousness. Focus, attention, vitality and mental clarity are all said to improve with beta training (Paul, 2008). The stress levels of Adults like, Professional course college students, and IT Employees, Doctors, Workers were shown to be lowered following yoga performance in this research.

CHAPTER 9
CONCLUSION

According to the National Crime Records Bureau of the Ministry of Home Affairs of the Government of India (2000), 2320 children every year, or more than six children per day, commit suicide as a result of failing exams. This shocking statistic demonstrates the gravity of the issue with pupils' stress levels. Young Indians who live in affluent and better-educated areas are more likely to face such issues.

Anxiety has a detrimental impact on the emotional and intellectual development of College students and Working Professionals. After doing yoga, the training technique and stress levels, both physical and mental, were steadily lowered and monitored. Depression has emerged as the most frequent health issue among university/college students, particularly among the Millennial Generation. Poor social problem-solving, cognitive distortions and family conflict are all possible causes of depression, as can estrangement from parents and peers, helpless attribution style, gender, and perceived instructor criticism. InIndia, mental health issues in children and adolescents are common, but they are rarely given the attention they need because many people are unaware thatmental illness is a leading cause of suicide. Approximately six Indian students commit themselves every day as a result of academic stress and test failure. We may deduce from prior research that Indian students are more prone to a lot of stress throughout their studies and that if they are exposed to excessive stress/frustration, they are more likely to commit suicide. Through yoga therapy, this study effort suggests a better diagnosis strategy for reducing stress, anxiety and depression in today's generation. According to this research continuous and repetitive surgical procedures generate modest mistakes during large operations, however after practicing yoga cognitive and surgical process errors are marginally decreased. The research also finds appropriate forms of yoga practice for lowering mental burden, cognitive function and as a result, suicidal thoughts in the participants.

When the amplitude of upper alpha increases as a result working memory performance improves. It was confirmed using the conceptual span test. This testwas done before and after the training and shows that working memory has improved. Alpha training is effective in boosting cognitive performance and lowering stress since it improves blood flow to the brain.

9.1 APPLICATION OF THE ALPHA TRAINING

- Cognitive deterioration as people become older
- Increases a depressed child's cognitive function
- used to address short-term memory problems.

9.2 FUTURE WORKS

This project's future study might encompass Alzheimer's illness, andAutism. This might be used to help those with short-term memory loss.

CHAPTER 10
REFERENCE

[1]. Datta, K., Bhutambare, A., VL, M., Narawade, Y. A., Srinath, R., & Kanitkar, M. (2023). Improved Sleep, Cognitive Processing and Enhanced Learning and Memory Task Accuracy with Yoga Nidra Practice in Novices. *medRxiv*, 2023-01.

[2]. León Murillo, J., Sotelo Monge, M. A., & Villagrá, V. (2023, February). Design of a Validation Model of the Cognitive State in Military Operations in Cyberspace. In *Computer Security. ESORICS 2022 International Workshops: CyberICPS 2022, SECPRE 2022, SPOSE 2022, CPS4CIP 2022, CDT&SECOMANE 2022, EIS 2022, and SecAssure 2022, Copenhagen, Denmark, September 26–30, 2022, Revised Selected Papers* (pp. 466-481). Cham: Springer International Publishing.

[3]. Sharpe, E., Butler, M. P., Clark-Stone, J., Soltanzadeh, R., Jindal, R., Hanes, D., & Bradley, R. (2023). A closer look at yoga nidra-early randomized sleep lab investigations. *Journal of Psychosomatic Research*, *166*, 111169.

[4]. Divarco, R., Ramasawmy, P., Petzke, F., & Antal, A. (2023). Stimulated brains and meditative minds: A systematic review on combining low intensity transcranial electrical stimulation and meditation in humans. *International Journal of Clinical and Health Psychology*, *23*(3), 100369.

[5]. Sánchez-Cifo, M. Á., Montero, F., & Lopez, M. T. (2023). A methodology for emotional intelligence testing in elderly people with low- cost EEG and PPG devices. *Journal of Ambient Intelligence and Humanized Computing*, 1-17.

[6]. Gunuc, S., & Koylu, E. (2023). Investigation of the Relationships between Beck Depression/Anxiety Scores and Neuropsychological Tests Scores with Lifestyle Behaviors in the Context of Neuroplasticity and Neurogenesis Approach. *Neuroscience*.

[7]. O'Hare, A. J., & Gemelli, Z. T. (2023). The effects of short interventions of focused-attention vs. self-compassion mindfulness meditation on undergraduate students: Evidence from self-report, classroom performance, and ERPs. *PloS one*, *18*(1), e0278826.

[8]. Burns, J. W., Jensen, M. P., Gerhart, J., Thorn, B. E., Lillis, T. A., Carmody, J., & Keefe, F. (2023). Cognitive therapy, mindfulness-based stress reduction, and behavior therapy for people with chronic low back pain: A comparative mechanisms study. *Journal of Consulting and Clinical Psychology*.

[9]. Kumar G, P., Sharma, K., Manvi, A., & Angarai, R. G. (2023). Coherence-based interhemispheric EEG functional connectivity changes in distinct frequency bands during

eyes open meditation. *bioRxiv*, 2023-02.

[10]. Kanna, R. K., & Vasuki, R. (2019). Advanced Study of ICA in EEG andSignal Acquisition using Mydaq and Lab view Application. *International Journal of Innovative Technology and Exploring Engineering (IJITEE) ISSN*, 2278-3075.

[11]. Olaniyan, O. T., Adetunji, C. O., Dare, A., Adeyomoye, O., Adeniyi, M. J., & Enoch, A. (2023). Neural signaling and communication using machine learning. In *Artificial Intelligence for Neurological Disorders* (pp. 245-260). Academic Press.

[12]. Varde, P. V. (2023). Human Factors in Operation. In *Risk-Conscious Operations Management: An Integrated Paradigm for Complex Engineering System* (pp. 281-366). Singapore: Springer Nature Singapore.

[13]. Kripa, N., Vasuki, R., & Kanna, R. K. (2019). Realtime neural interface controlled au-pair BIMA bot. *International Journal of Recent Technology and Engineering*, *8*(1), 992-4.

[14]. Pellegrino, M., Ben-Soussan, T. D., & Paoletti, P. (2023). A Scoping Review on Movement, Neurobiology and Functional Deficits in Dyslexia: Suggestions for a Three-Fold Integrated Perspective. *International Journalof Environmental Research and Public Health*, *20*(4), 3315.

[15]. Kripa, N., Vasuki, R., & Kanna, R. K. Advanced Approach for HyperActivity Children Treatment Using EEG and EMG Biofeedback System. *International Journal of Psychosocial Rehabilitation*, *24*(8), 5423-6.

[16]. Rodrigues, J., Matos, R., Jacinto, M., & Rodrigues, F. (2023). Book ofabstracts II International Congress-CIEQV 23rd and 24th February 2023.
In *II International Congress-CIEQV*. Centro de Investigação em Qualidade de Vida-CIEQV, Escola Supeior de Educação e Ciências Socias-Politécnico de Leiria.

[17]. Kanna, R. K., Kripa, N., & Vasuki, R. Systematic Design Of Lie Detector System Utilizing EEG Signals Acquisition. *International Journal ofScientific & Technology Research*, *9*, 610-2.

[18]. Marsh, E. B., Girgenti, S., Llinas, E. J., & Brunson, A. O. (2023). Outcomes in Patients with Minor Stroke: Diagnosis and Management in the Post-thrombectomy Era. *Neurotherapeutics*, 1-12.

[19]. Mohapatra, S. K., Kanna, R. K., Arora, G., Sarangi, P. K., Mohanty, J., & Sahu, P. (2022, October). Systematic Stress Detection in CNN Application. In *2022 10th International Conference on Reliability, Infocom Technologies and Optimization (Trends and Future Directions)(ICRITO)* (pp. 1-4). IEEE.

[20]. Oliver IV, C. (2023). *The Effects of Strength Training on Psychological Functioning in Adolescents* (Doctoral dissertation, Michigan School of Psychology).

[21]. Kanna, K., Prasath, T. M., & Vasuki, R. (2019). Design of a Light Activated Switch for Persons with Cerebral Palsy. *Indian Journal of Public Health Research & Development, 10*(11).

[22]. Chadman, K. K., Keller, R., Srinivasan, S. M., Amonkar, N., Su, W. C., Bhat, A. N., & Behavioral-Affective, S. (2023). Edited by: Cheryl Dissanayake, La Trobe University, Australia Reviewed by. *Early intervention in autism spectrum disorder (ASD), 16648714.*

[23]. Kanna, R. K., Prasath, T. M., Subhalakshmi, N., & Vasuki, R. (2019). Intelligent helmet for bikers using sensors. *Drug Invention Today, 11*(7).

[24]. Kala, N., Telles, S., Sharma, S. K., & Balkrishna, A. (2023). P300 Following Four Voluntarily Regulated Yoga Breathing Practices and Breath Awareness. *Clinical EEG and Neuroscience, 54*(2), 117-129.

[25]. Kanna, R. K., Prasath, T. M., & Vasuki, R. (2019). Performance Analysis of FPGA Based Fund Us Image Processing. *Indian Journal of Public Health Research & Development, 10*(11).

[26]. Pandit, S. A. (2023). *Decolonizing Consciousness: Reclaiming the Indian Psychology of Well-being.* Taylor & Francis.

[27]. Kanna, R. K., Banappagoudar, S. B., Sundararajan, V., shalini Joseph, P., Jalal, S. M., & Dash, B. (2022). MACHINE LEARNING BASED DRONE DESIGN FOR COVID-19APPLICATION. *NEUROQUANTOLOGY, 20*(15), 2039-2047.

[28]. Banushi, B., Brendle, M., Ragnhildstveit, A., Murphy, T., Moore, C., Egberts, J., & Robison, R. (2023). Breathwork Interventions for Adults with Clinically Diagnosed Anxiety Disorders: A Scoping Review. *Brain Sciences, 13*(2), 256.

[29]. Kanna, R. K., Gomalavalli, R., Priya, C., & Gunasekaran, K. (2023). DETECTION OF MENTAL FATIGUE IN DRIVERS USING EEG APPLICATIONS. *Journal of Data Acquisition and Processing, 38*(1), 343.

[30]. Premkumar, N. K., & Vasuki, D. R. (2016). Extraction and analysis of EEG waves for the study of enhancement and rejuvenation of brainwaves through meditation. *Int J Pharma Technol, 8*, 13481-8.

Publisher: Eliva Press Global Ltd

Email: info@elivapress.com

Made in United States
Orlando, FL
04 November 2026

90200864R00036